WCMC Handbooks on Biodiversity Information Management

Volume 1 Information and Policy

World Conservation Monitoring Centre

WORLD CONSERVATION
MONITORING CENTRE

Series Editor J.H. Reynolds

Commonwealth Secretariat 1998

The **World Conservation Monitoring Centre**, based in Cambridge, UK, is a joint venture between three partners in the World Conservation Strategy and its successor Caring for the Earth: IUCN – The World Conservation Union, UNEP – United Nations Environment Programme, and WWF – World Wide Fund for Nature. The Centre provides information services on the conservation and sustainable use of species and ecosystems and supports others in the development of their own information systems.

The United Kingdom's **Darwin Initiative for the Survival of Species**, launched at the 1992 Earth Summit in Rio de Janeiro, aims to support the Convention on Biological Diversity by drawing on Britain's scientific, educational and commercial strengths to assist in the conservation and sustainable use of the world's biodiversity and natural habitats. Key tenets of the Darwin Initiative include collaboration and cooperation with local people, capacity building, distinctiveness and complementarity of project initiatives, poverty alleviation, and long-term sustainability. Through training, awareness raising, and research on undervalued areas of biodiversity, Darwin support is particularly aimed at strengthening links between Britain and those countries rich in biodiversity but poor in financial resources.

Under the auspices of its **Environmental Training for Sustainable Development** initiative, the Management and Training Services Division of the **Commonwealth Secretariat** supports short- and long-term training, internships and institution development for environmental policy makers, environmental 'operatives', and environmental information professionals in the Commonwealth, in various areas of the environment including biodiversity and gender. Funding support for training, institution development and publications under the aegis of the Management and Training Services Division is provided by the Fund for Technical Co-operation (CFTC).

WORLD CONSERVATION MONITORING CENTRE

Published by Commonwealth Secretariat

ISBN 0-85092-544-4

Citation World Conservation Monitoring Centre. 1998. *WCMC Handbooks on Biodiversity Information Management. Volume 1: Information and Policy*. Reynolds, J.H. (Series Editor). Commonwealth Secretariat, London. ix + 29pp.

Typeset by Bookcraft Ltd, Stroud, Gloucestershire, England

Cover design Michael Edwards

Photography by J.S. Donaldson (Wood's cycad, *Encephalartos woodii*); D. & I. Gordon (Mali landscape; Plant study, Ghana; Thai forest; Rock hyrax, *Procavia capensis*); IUCN/J. McEachern (Diver and fish); WCMC (Ecoregion and Africa maps; GIS work; Workshop facilitation).

Available from IUCN Publications Services Unit
219 Huntingdon Road, Cambridge, CB3 0DL, UK
Tel: +44 1223 277894; Fax: +44 1223 277175
Email: iucn-psu@wcmc.org.uk
http://www.iucn.org/bookstore/

CONTENTS

ACKNOWLEDGEMENTS

The generous support of the *United Kingdom's Darwin Initiative for the Survival of Species* has provided for the development of a comprehensive programme of training in biodiversity information management. This programme comprises an international training team, drawing on expertise from collaborating organisations around the world; the preparation of a training resource in the form of a handbook series and related materials; and the development of computer-based demonstration tools. Training is being promoted through the delivery of post-graduate modules, and through regional and national workshops which have received additional support from The British Council, British Airways Assisting Conservation Scheme, and contributions from participating organisations. The programme has been appropriately titled *Darwin Initiative Training in Biodiversity Information Management.*

Development of the handbooks has also benefited from experiences gained through the Biodiversity Data Management (BDM) Project, administered by the United Nations Environment Programme (UNEP) and funded by the Global Environment Facility (GEF), and related initiatives supported through the European Union (EU) and European Environment Agency (EEA). Indeed, Volume 6 draws extensively on one of the key outputs of the BDM Project, the *Guide to National Institutional Survey* (UNEP/WCMC 1998), developed in consultation with participating countries, the BDM Advisory Committee and the UNEP management team. The concept of an information cycle was developed in collaboration with the International Institute for Environment and Development (IIED) with support from the UK Department for International Development (DFID). The handbooks have been published through the generous support of the Commonwealth Secretariat.

Fundamental to the development of this programme have been the partnerships established with training organisations around the world. These organisations have worked collaboratively in hosting workshops, in reviewing the handbook materials, and in providing guidance on how regional and national training needs can be met most effectively. The training programme has significantly benefited from the input of numerous individuals working in the field of biodiversity information management. Among these individuals, particular mention goes to Professor Ian Crain and Gwynneth Martin of the Orbis Institute, Ottawa, Claire Appleby, an independent consultant, and to Drs Jake Reynolds and John Busby of WCMC for their insightful work in developing the handbook series. Thanks are also extended to Laura Battlebury for her tireless administrative and logistical support. The series editor for the handbooks was Jake Reynolds, while Donald Gordon managed the overall project.

To the many individuals, both within and outside WCMC who have contributed to the development of materials and the delivery of training in biodiversity information management, a profound debt of gratitude is owed. It is through this collaborative effort that a service is being developed to contribute to the conservation and sustainable use of living resources.

BACKGROUND

The purpose of the *WCMC Handbooks on Biodiversity Information Management* is to support those making decisions on the conservation and sustainable use of living resources. The handbooks form part of a comprehensive programme of training materials designed to build information-management capacity, improve decision-making and assist countries in meeting their obligations under Agenda 21 and the Convention on Biological Diversity.

The intended audience includes information professionals, policy-makers, and senior managers in government, the private sector and wider society, all of whom have a stake in the use or management of living resources. Although written to address the specific need for improved management of biodiversity-related information at the national level, the underlying principles apply to environmental information in general, and to decision-making at all levels. The issues and concepts presented may also be applied in the context of specific sectors, such as forestry, agriculture and wildlife management.

The handbooks deal with a range of issues and processes relevant to the use of information in decision-making, including the strengthening of organisations and organisational linkages, data custodianship and management, and the development of infrastructure to support data and information exchange. Experience suggests that some of the greatest challenges in information management today are concerned with organisational issues, rather than technical concerns in the delivery of information which supports informed decision-making. Consequently, topics are addressed at management and strategic levels, rather than from a technical or methodological standpoint, and alternative approaches are suggested from which a selection or adaptation can be made which best suits local conditions. Nevertheless, in adopting this framework approach, we have tried to adhere to recognised conventions and formalisms used in information management and trust that in producing a 'readable' set of handbooks the integrity of the materials has not been compromised.

Overall, the handbook series comprises:

Companion Volume	
Volume 1	*Information and Policy*
Volume 2	Information Needs Analysis
Volume 3	Information Product Design
Volume 4	Information Networks
Volume 5	Data Custodianship and Access
Volume 6	Information Management Capacity
Volume 7	Data Management Fundamentals

Collectively, the handbook series promotes a shift from tactically based information systems, aimed at delivering products for individual project initiatives, to strategic systems which promote the building of capacity within organisations and networks. This approach not only encourages data to be managed more effectively within organisations, but also encourages data to be shared amongst organisations for the development of the integrated products and services needed to address complex and far-reaching environmental issues.

The handbook series can be used in a number of ways. Individual handbooks can be used to guide managers on specific aspects of information management; they can be used collectively as a reference source for strategic planning and project development; they can also provide the basis for a series of short courses and training seminars on key challenges in information management.

The companion volume provides the background to the handbook series. It also assists readers in deciding which handbooks are most relevant to their own priorities for strengthening capacity.

A second series of handbooks is planned to provide more detailed guidance on information management methodologies, including the areas of data and technology standards, database design and development, application of geographic information systems (GIS), catalogues and metadatabases, and the development of decision-support systems. The current series deals only briefly with formal system development methodologies, and for more detailed treatments the reader is encouraged to access the wide range of published and electronic resources available in libraries and on the Internet, some of which are alluded to in individual handbooks and reference sections.

A number of computer-based training tools have been developed to accompany the handbook series and are used in the training programme. These are based on a protected areas database, a tree conservation database, a GIS demonstration tool and a metadata directory. They aim to demonstrate key aspects in the collection, management and analysis of biodiversity data, and the subsequent production and delivery of information. They also illustrate practical issues such as data standards, data quality-assurance, data access, and documentation. Each training tool is supported by a user guide, together with a descriptive manual which traces the evolution of the tool from design, through development to use.

1 INTRODUCTION

Visions of degraded landscapes, changing climates, chronic pollution and extinction of species have permanently altered our view that the Earth is a limitless resource for exploitation. We now realise that our social and economic goals cannot be pursued regardless of their impact on the environment. Serious challenges lay ahead in translating such awareness into action. In particular, how can we progress towards sustainable development when the gap between current practices and sustainability appears to be so immense?

Many nations are responding to this challenge by committing to the principles of **Agenda 21**, the key output of the Earth Summit in 1992. In particular, they have opted to become Parties to international agreements and protocols relating to biodiversity conservation (e.g. Convention on Biological Diversity, CITES, Convention on Migratory Species, Ramsar Convention, World Heritage Convention) and stabilisation of the global environment (e.g. Framework Convention on Climate Change, Montreal Protocol). Five years later in 1997, at Earth Summit II, there was general agreement that enough discussion, negotiating and treaty-writing had taken place, and that the next major step forward would be **implementation**.

Central to the implementation process is the formulation of **policies, strategies** and **action plans** to promote the conservation and sustainable use of living resources. The performance of these policies depends on the degree to which they reflect the perspectives of different groups of people and, consequently, on the extent to which government, the private sector and society at large work in partnership. Experience has shown that the best policies result from a process of consultation, consensus-building and, if need be, reconciliation amongst affected stakeholders, leading to solutions which balance economic and social goals with the need to safeguard the environment. When pursued intelligently, a commitment to such policies can improve the economic performance of companies and allow longer-lasting benefits to be delivered by governments.

Good policies have other features in common, including the effective use of information throughout the policy's lifetime. Indeed, the transition from exploitation of living resources to conservation and sustainable use requires a major investment in information and monitoring, otherwise it is not possible to say whether we have truly progressed.

2 INFORMATION PRIORITIES

Amongst all the potential uses of information in addressing environment concerns, it may be argued that policy-development is the most important, since so many human activities are influenced by the policies of governments, corporations and, increasingly, non-governmental organisations. Environmental information, therefore, needs to be relevant to policy-makers in these sectors who are charged with identifying environmental concerns and finding solutions in complex and diverse situations, such as industries and communities.

Not surprisingly, many environmental concerns which affect the lives of ordinary people are not officially acknowledged in government policies, which may lag behind other sectors in society, such as lobbying groups, or be provided with too few resources to address the issues effectively. In the present context, the wider set of environmental concerns (which may or may not be acknowledged by governments and other groups) will be referred to as **policy issues**. In any particular location, factors such as culture, history, trade, politics, climate, geographic and biological composition affect which policy issues will be most relevant. Box 1 lists a few common issues to illustrate something of their breadth. A key step forward is to establish which issues are highest priority.

Box 1 Common policy issues

- Conversion of natural landscapes (e.g. forest to agriculture or housing).
- Decline in commercially valuable species (e.g. timber trees, wild foods and medicines).
- Degradation of ecosystems (e.g. loss of species diversity or ecosystem services).
- Loss of genetic variability (e.g. wild ancestors of crops).
- Release of exotic or genetically modified organisms (e.g. weeds and pests).
- Loss of indigenous knowledge (e.g. traditional forest-related knowledge).

Stakeholders may attribute widely differing priorities to policy issues according to their own individual perspectives. For instance, global warming could be one of the major driving forces of environmental change in the next century, but it is likely to receive scant attention in a low-income community challenged with day-to-day survival. Priorities also change over time: they arise, come to the attention of specific groups, and disappear, often to resurface later in a different guise.

The key challenge for information professionals is to understand not only **what** information is relevant at a specific time, but also **when, how and to whom** it should be delivered, in order to address important issues effectively.

Recognising that financial resources for information management are limited, it is vital for information professionals to work alongside other stakeholders in identifying priority areas for information development. To ensure that the interests of an appropriate range of stakeholders are represented — whether these are information specialists, policy-makers, resource users or other groups — a high degree of consultation is desirable during this process. Indeed, the very act of discussing priorities can build important ties between stakeholders as well as sensitising them to the role and value of environmental information in their daily work.

One challenge is the **scope of the issues** under consideration. For example, it would be difficult to prepare comprehensive information on complex topics such as 'poverty', 'population growth' or 'deforestation', since the range of perspectives on these issues (and consequently solutions) is enormous. Faster progress can be achieved by generating information on more focused issues such as 'loss of breeding habitat for species X due to drainage of wetland Y' or 'effect of river pollution on species Z'. The challenge is to be realistic about what can be achieved with the resources available, and prioritise accordingly.

Another factor is the **lifetime of the issues** under consideration. At one extreme, it is difficult to generate information on so-called 'burning issues', such as the effects of an imminent road-building project, since they remain topical only for unpredictable, and usually short, periods of time. At the other extreme, research on 'fundamental issues', for example the effect of climate change on wild bird populations, may arrive too late to save many of the species under study.

It is desirable to forecast issues well before they develop into problems, so that cost-effective remedial action can be organised. The best way of achieving this is to manage (or have access to) data on topics which yield information on immediate priorities, but are also relevant to the study of longer-range phenomena. The capacity to 'scan the horizon' for future issues, using a combination of historical experience and up-to-date intelligence, is a hallmark of effective information management.

3 PARTICIPATION AND CONSENSUS

One of the reasons why environmental issues are so challenging is that the values attached to natural resources vary tremendously between stakeholders. In some cases values may have evolved over thousands of years of interaction with the environment and be embedded in the spiritual life or culture of a community. In other cases they may have been acquired recently in response to new economic pressures or opportunities. A paradox exists since, in many cases, the signs of environmental deterioration are clear to many, yet there is no consensus on how to respond.

Traditionally, economic goals have carried most sway over development decisions, as depicted in Figure 1. One of the reasons for this is that social and environmental values are more difficult to quantify than economic values. For instance, how does one quantify the potential value of an undiscovered species to the pharmaceutical industry, or the detriment to society of a species extinction? Such uncertainty has the effect of encouraging stakeholders to underestimate environmental values in their decision-making. For this reason, large-scale projects may be subjected to social and environmental impact assessments (EIA) before they are implemented, ostensibly to ensure that steps have been taken to minimise non-economic costs. However, confidence in this process is still low in many

Figure 1 Changing visions of development: the search for sustainability

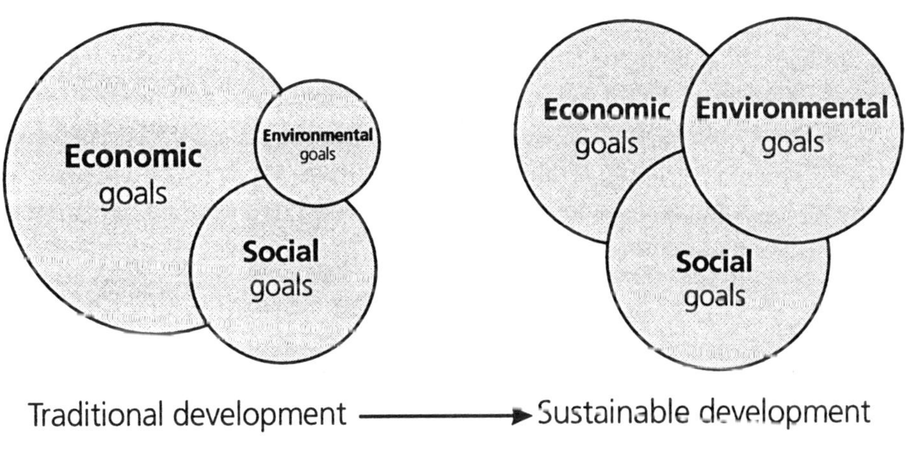

countries where such assessments have not been successful in preventing social and environmental damage.

Now that the costs of economic development are becoming more widely understood, world attention is being focused on an alternative vision — known as **sustainable development** — in which economic goals are achieved in balance with social or environmental goals (see Figure 1). This may not satisfy all stakeholders (e.g. those who would wish to see more rapid economic development), but most may expect to benefit in the long term and, critically, the benefits of development will be shared more equally.

Individuals, communities, companies, nations and international bodies all make decisions which affect the sustainability of living resources and, consequently, all have a role to play in their conservation and sustainable use. No segment or level of society can be left out, since decisions made by one group can affect the livelihoods of others. Thus, wherever possible, policies aiming to conserve living resources should reflect the perspectives and needs of all stakeholders who stand to win or lose. Even small-scale environmental challenges tend to arouse the interest of many stakeholder groups. Such groups are typically politicians, civil servants, natural resource managers, local government officials, non-governmental and community-based organisations, business leaders, industry representatives, professional associations, scientific researchers, teachers, the general public, media and the international community.

Despite their apparent diversity, most of those involved fall into one of three categories: **government, private sector** and **civic society**. Together, these are referred to as the development 'triad' (Sandbrook 1994), since they represent the three core interests shaping development policy. In the long term, policies which do not represent all three broad interests are destined to falter, stall or fail. Figure 2 depicts the development triad in terms of its constituent stakeholders. Clearly, the term 'civic society' represents all of those who are not actively engaged with government or commercial activities, including non-governmental organisations, community-based organisations and individuals.

Figure 2 The development triad

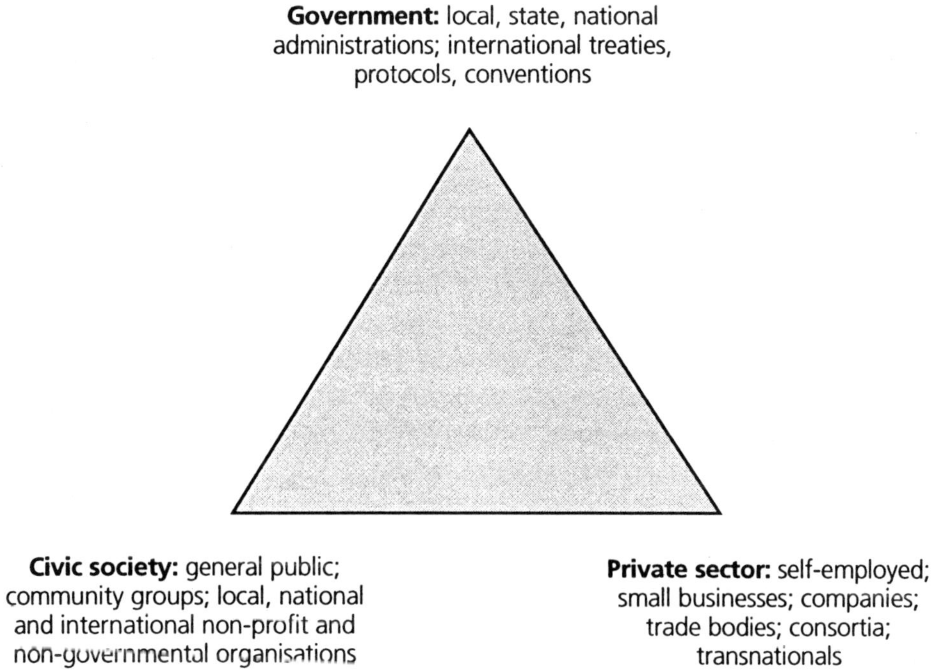

4 SUPPORTING GOOD POLICIES

For governments to promote the conservation and sustainable use of living resources, a **credible policy environment** is required, i.e. one in which the planning, regulatory and advisory services provided by the government are understood and accepted by the majority of stakeholders. This is best achieved by involving stakeholders in the policy-development process, by steadily adapting policies in response to feedback from stakeholders, and by reviewing policy effectiveness on a regular basis.

Credible policies on biodiversity conservation are required within all levels and sectors of government, and outside government where industries, firms, and communities also have a responsibility to reduce environmental impacts and protect (or sustainably use) living resources. The important parallel between national and local levels, or governmental, private and non-governmental sectors, is that policy-development is an inclusive, participatory activity, depending on high-quality and objective information for success. In this way, decision-making processes are not only made more transparent, **they are also shared**.

Several models have been proposed for biodiversity planning — the development of policies, legislation, strategies and action plans for conservation and sustainable use of living resources (e.g. UNEP 1993, Miller and Lanou 1995). Good information management is assumed in these models, but its precise role, for instance in delivering efficiency gains and cost savings, is not always elaborated. Figure 3 illustrates a generic management 'loop' comprising four processes: **plan, implement, monitor** and **review**. By making extensive use of information, the four processes enable policy goals to be achieved in a progressive manner through successive iterations of the loop.

Joining the loop at the implementation stage, activities are underway to meet agreed targets for conservation and sustainable use, set out in the planning stage. The loop proceeds to monitor the performance of the policy by, for example, obtaining data directly from the environment or with respect to the achievement of policy targets. Performance is then reviewed, leading to the production of clear and concise recommendations for policy-makers on how the policy should be refined in future. Finally, the loop is 'closed' for another cycle by planning how the recommendations will be implemented, in terms of objectives, targets, roles and responsibilities. It should be noted that monitoring, review and planning activities would very often proceed in parallel with implementation. This enables continuous, rather than intermittent, feedback and policy-refinement.

Figure 3 Management loop for policy-development

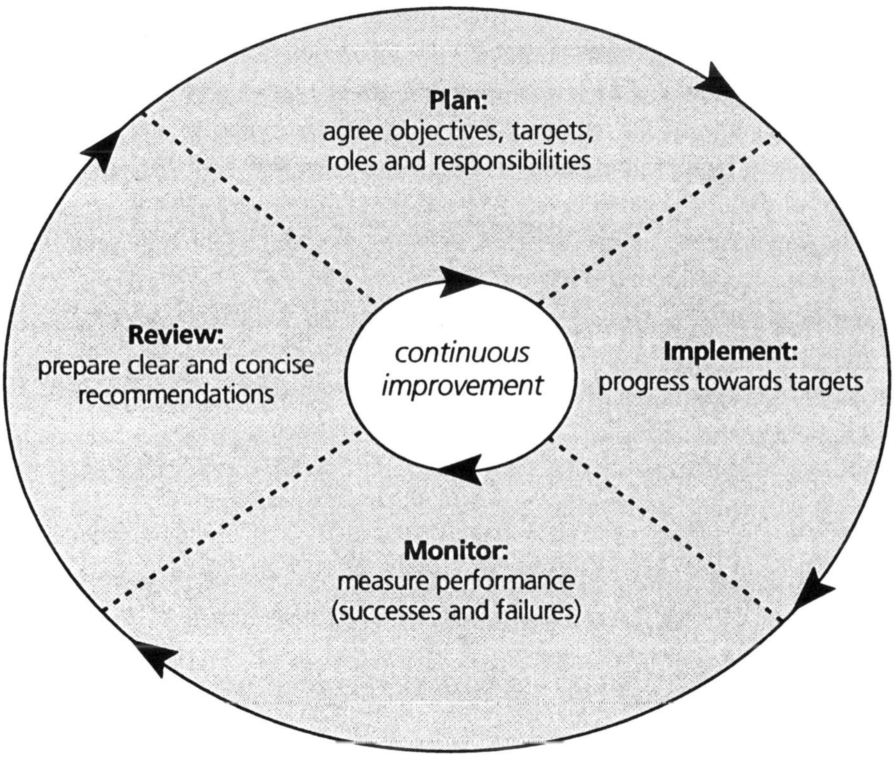

Figure 3 simplifies what, in reality, is a complex, many-faceted process. For instance, the policy being developed may address an issue of national importance, such as the loss of crop genetic variability in an important agricultural zone, or a local concern, such as the restoration of a single eroded hillside. In both cases successful implementation depends on maintaining a steady course around the management loop so that the four components flow into one another. To achieve this it is necessary to concentrate on the core objectives of the policy at all stages and to make sure that **all** actions contribute to these in some way.

An open, participatory approach encourages stakeholders from different levels and sectors of society to involve themselves in implementing the management loop. For instance, policy goals may be developed by consensus at fora established for this purpose; community groups or industry representatives can be asked to help monitor policy performance; and the success or failure of policies, plus options for future refinement, can be determined as a group.

When stakeholder perspectives differ greatly, participation may not always be easy to manage. In such cases considerable effort is needed to keep stakeholders focused on policy goals, and not let implementation be diffused. Common constraints on implementation include lack of financial resources, expertise and time, political interference and lack of political will. Clearly, the management loop is not an easy approach to follow, but it does offer a powerful means for developing and implementing policies capable of effectively responding to changing environmental conditions in a controlled, transparent fashion.

5 THE DECISION-MAKING PROCESS

When wrestling with problems like deforestation, water pollution, loss of biodiversity and a fragile food supply, decision-makers are presented with a series of major challenges relating to information. Firstly, they must learn how to **discriminate between information sources**, which may carry different messages and vary in quality. Secondly, they must learn how to **integrate different types of information** to formulate a proper perspective on the issues concerned. Finally, they need to clarify information into a series of **practical options** which may be evaluated. To address these challenges fully may require decision-makers to liaise effectively with the scientific community and make appropriate use of information technology.

The unprecedented amount of information which is now available to decision-makers should help them arrive at sounder, faster and more transparent decisions. However, these expected benefits often do not materialise in practice. One reason is that although the amount of environmental information has grown, so has the apparent complexity of the issues which need to be tackled. This is most clearly evidenced in terms of the numbers of stakeholders who now expect to be involved in environment-related decisions, and the challenges this creates in finding a consensus.

Realistically, environmental information is only one factor influencing the way in which decisions on living resources are made, and it is not always the most significant. Personal opinions, political and economic dogma, market pressures, and legal necessities also influence decision-making. However, information has the power to be more objective than other influences, especially when generated by recognised means such as the scientific method. The importance of information to decision-making was noted at the Earth Summit in 1992, in the form of a whole chapter of **Agenda 21** (Chapter 40). The **Convention on Biological Diversity** also places information management as a top priority (see Articles 7 and 17), notably in connection with the preparation of national biodiversity strategies and action plans (Article 6), and reports on measures taken to implement the Convention's goals (Article 26).

The establishment of a modern theory of decision-making is often attributed to Herbert Simon (see for instance Newall *et al.* 1958) who postulated a three-stage process: intelligence, design and choice. In the intelligence stage the decision-maker gathers information relevant to the subject or problem; in the design stage alternative courses of action (options) are identified and their consequences forecast on the basis of available information; finally a preferred course of action is chosen. The decision-making process is considered to be 'rational' if the choice is based on objective criteria, such as the minimisation of costs or the maximisation of benefits.

Whether truly rational or not, whether supported by computer systems or not, the decision-making process will usually be similar to that shown in Figure 4 (adapted from Brookes *et al.* 1982). Important elements are the information sources, which enable options to be evaluated objectively, and a feedback loop to ensure that the impact of a decision is both monitored and reviewed. The rational argument for conservation and sustainable use of living resources is often very clear. What remains is to raise awareness of this argument over competing interests, so that the 'gut' feelings of decision-makers are influenced by objective, not subjective criteria.

Brookes *et al.* (1982) point out that rationality becomes difficult when subjective concepts such as 'quality of life' and other cultural values influence the decision-making process. In such cases, there is a risk that rational options may be ruled out before they have been properly considered, by virtue of their cultural or personal unacceptability. Of course, subjective influences do not always work in opposition to the environment. Many indigenous peoples, for example, have belief systems which are naturally sympathetic to conservation goals.

Figure 4 The decision-making process

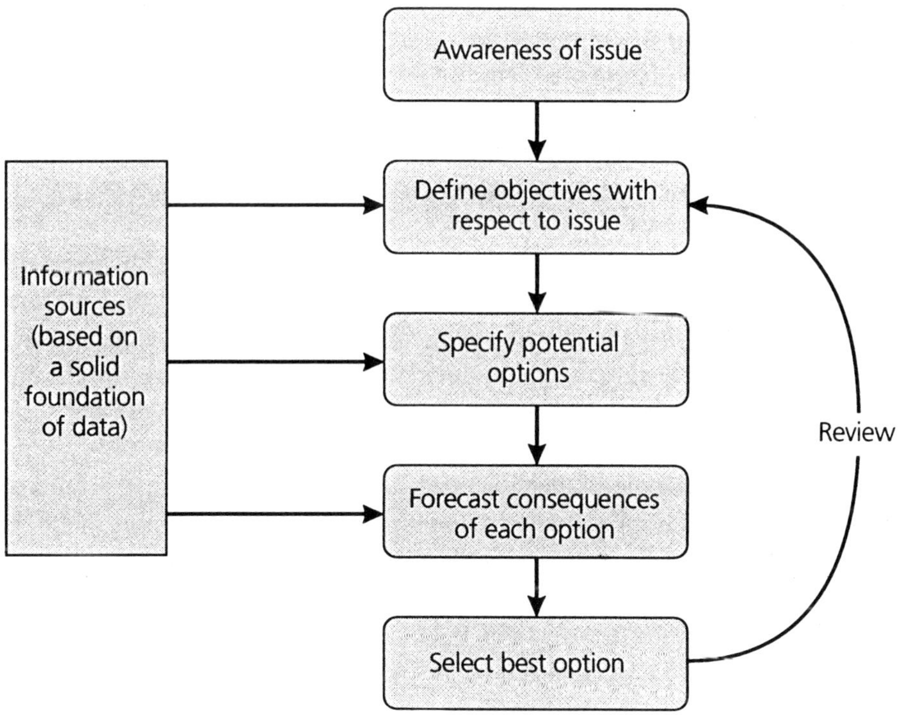

6 SUPPORTING GOOD DECISIONS

The fact that information is available does not guarantee that it will be used. In fact, no benefit may be obtained unless there is an effective dialogue between those involved in the process of generating the information (e.g. scientists, researchers and other information professionals) and those who are expected to use it (e.g. policy-makers and planners). For a variety of reasons this dialogue is often absent or under-developed. The existence of long-standing barriers between scientists and policy-makers, reinforced by arcane scientific literature and ill-informed policies, is one of the contributing factors.

Clearly, information may fail to achieve a significant impact on decision-making when it is developed **outside** of mainstream decision-making processes, rather than **emerging from within**. For example, a local wildlife society may record the decline in breeding turtles on an increasingly polluted stretch of beach year after year. To raise awareness of the crisis they may hold meetings, write stories in the press, publish newsletters and alert other societies, scientists and concerned groups to the issue. However, until the government commissions its own enquiry into the problem, appropriate and decisive action may not be taken.

Disclosing information to the general public can succeed in provoking government (or other elements of society) into action,[1] but the way in which the government reacts depends on the internal information it has at its disposal. An understanding of the political climate and cultural values of the country is necessary before deciding which is the most effective means of conveying information to decision-makers. In the above example, a better strategy for the wildlife society may be to seek a working partnership with the government, rather than embarrassing it, with the aim of adjusting policy well ahead of impending crises (this assumes that the government in question is willing to listen and seek advice from wider elements of society).

Organisations of all kinds, whether governmental, commercial or community-based, are driven by their own priorities and needs. Thus, in the midst of a complex privatisation process, under external pressure to fulfil key targets within a structural adjustment programme, it may be difficult, even impossible, to interest senior

---

1 This is known as 'decision-making by disclosure' — influencing policy-makers via the public domain, rather than by more bureaucratic channels.

governmental decision-makers in important environmental information. At best they may see such information as a distraction and, more probably, as a waste of precious time. This helps to explain why information generated outside of decision-making circles is often ignored, misunderstood or viewed as a threat, and this does not only apply to governments!

For information on environmental issues to be appreciated, those who are expected to use it need to be aware of how and why it has been produced and, preferably, **have been involved** in some way in its production. Thus, it is important to emphasise the role of environmental information within the day-to-day management contexts of those organisations which have the most impact on living resources. These include stakeholders in government, the private sector and civic society, which have a direct role in managing or using living resources. They also include financial institutions, the media and other sectors which, due to their power, are capable of great influence over attitudes and policy in other areas.

An important justification for increased attention to information management is that the contexts in which policies on living resources are formulated are usually complex, and **do not respond well to simple solutions**. Many viewpoints may have to be reconciled to find the most appropriate way forward, a process which depends on building consensus between stakeholders on the basis of mutual respect and objectively produced information. Information management is also central to performance monitoring, whereby the success (or otherwise) of policy solutions is fed back to enable more effective policies to be designed in future. Review mechanisms of this kind are a standard feature of many economic policies but, in the case of conservation, the expertise and resources needed to collect, analyse and review environmental data are often underestimated.

Properly organised information brings transparency and objectivity into the process of decision-making, throwing light on complex issues and providing a means of comparing potential solutions. Specifically, good information empowers its audience by:

- providing a range of options on which to base a decision;

- discouraging options with predictably adverse consequences; and

- adding to a common set of agreed facts on which to base discussion

Information rarely achieves these goals unless it is relevant to the needs of its users, is timely (i.e. available when and where it is needed), and can be easily absorbed without the need for special training or technology. For example, information published in a scientific journal serves academics and researchers well, but its presentation and timing may be unsuitable for policy-makers who are focused on operational goals. Similarly, raw data on the status of endangered species may not be usable by busy managers without significant interpretation or integration with other data.

In short, decision-making is most effective when it is based on **insight**, not simply additional sources of information or data. The distinction between information and insight is difficult to draw but, in general, the latter is characterised by a more rounded product, tailored to the needs of the decisions being taken, and very concise (see Volumes 2 and 3). In the case of decisions on living resources, a package of information describing the state of a particular resource, the pressures it faces from human activities, and the effectiveness of current policy and regulatory measures designed to conserve it, could claim to offer a degree of insight beyond what would be provided by information on any of these topics individually.

7 THE INFORMATION CYCLE

Many aspects of the management loop introduced in Section 4, particularly the planning, monitoring and reviewing processes, depend on access to high-quality objective information on a potentially diverse set of topics. This provides a rather daunting challenge for information professionals, who may find themselves able to deal with specific themes, but totally unable to marshal the full range of information sources. Indeed, an impressive range of inputs is required to manage information on any specific theme, including the underlying datasets, human resources, facilities and management systems. **Multi-disciplinary cooperation** is clearly the way forward for successful information production. To help those involved cooperate in a coherent, cost-effective, manner, a guiding framework is useful to focus their contributions on common objectives and targets.

Recognising that each country will respond uniquely to the challenge of generating information on environmental issues, a flexible, process-oriented approach is proposed. This breaks down the challenge into a series of small steps which, when taken as a whole, lead to the achievement of major strategic objectives. Figure 5 illustrates how the approach — referred to as the **information cycle** — covers all the steps necessary to address policy issues in a planned, yet responsive manner. The information cycle is not rigid, nor does it prescribe a solution: variable ordering of the processes is likely due to the completion of certain tasks, the need to revisit processes, or further specific constraints. Where individual processes in the cycle have already been accomplished, the cycle can still be used to suggest next steps or draw attention to missed, or under-emphasised activities.

The information cycle can be applied at a range of scales. It could work equally well within a single organisation, a group of organisations, or in a large network of organisations spanning multiple sectors, levels and disciplines (see Volume 4). One thing is certain: at whatever scale the information cycle is applied, a high degree of coordination is required to ensure that stakeholders are kept informed of key results, objectives and responsibilities. An effective way of keeping the information cycle on track is to establish an organisational structure to oversee its implementation. This could be an existing committee, steering group or other body with appropriate representation, or one specially created for the task. Ideally, the body should be closely, if not directly, involved in the policy-development process to ensure that the information cycle is anchored in decision-making machinery.

Figure 5 The information cycle

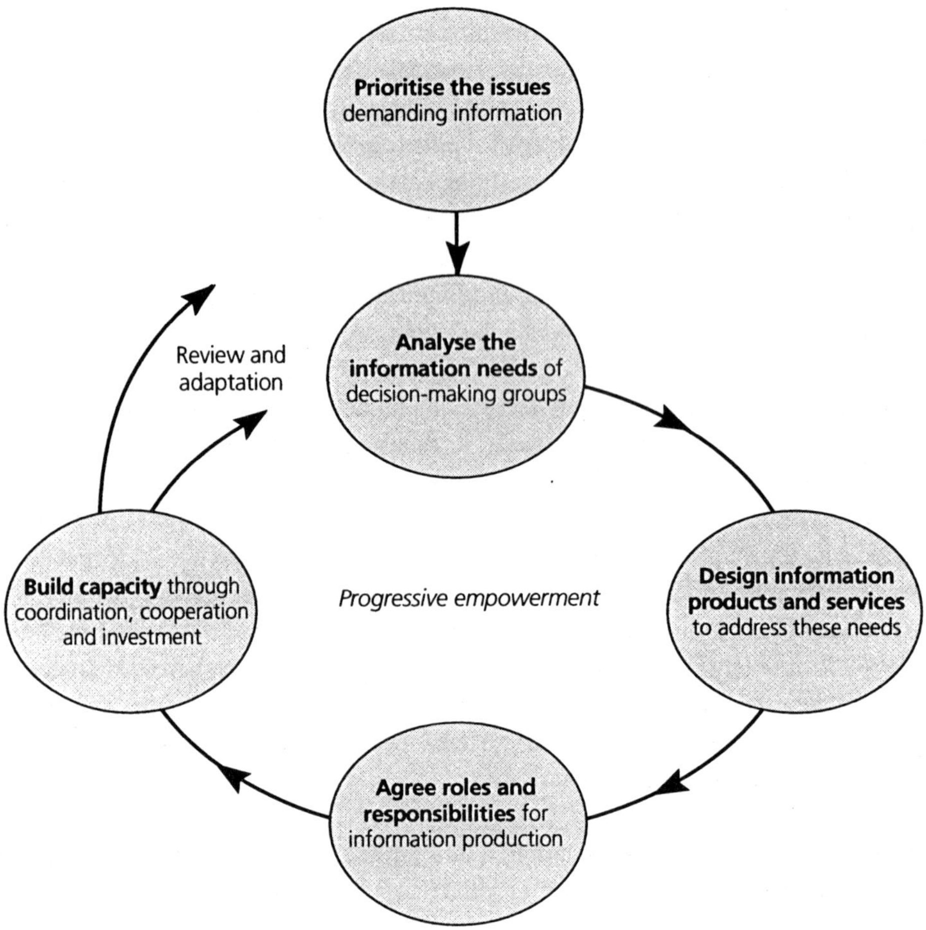

As an example of how the steering group might function, the prioritisation process could be facilitated by forming a working group to draft a list of priority environmental issues for which better information is required. These could be distributed to key stakeholders for review, following which a visioning exercise could be held to develop a consensus on their relative importance and information needs (see Volume 2).

The information cycle consists of five processes, one of which — **prioritise the issues** — should be undertaken before the others since it establishes which issues are the highest priority for information support and, therefore, which should be addressed first. The remaining four processes cover the activities necessary to produce cost-effective information on the issues selected. Recognising that priorities change over time, and that information needs will change with them, a secondary feature of the information cycle is **review and adaptation**, whereby progress through the cycle is reviewed and new targets defined as appropriate.

8 RECOMMENDED PROCESSES

Each individual process recommended in the information cycle is elaborated below.

1. Prioritise the issues

Recognising that resources for information management are limited, it is vital for information professionals to work alongside other stakeholders in identifying priority areas for information development. To ensure that the interests of an appropriate range of stakeholders are represented — whether these are information specialists, policy-makers, resource users or other groups — a high degree of consultation is desirable during this process. Indeed, the very act of discussing priorities can build important ties between stakeholders as well as sensitising them to the role and value of environmental information in their daily work. Key challenges to be aware of are the scope of the issues under consideration (i.e. complex versus focused) and their lifetime (i.e. 'burning' versus 'fundamental').

Further reading: this handbook.

2. Analyse information needs

Solutions to environmental issues are usually complex and it is not always obvious how to determine what information is needed to support policy and management goals. This is particularly true when decision-makers have only a hazy idea of their requirements. However, the price for not pursuing this challenge is heavy. Without the 'right' information, there is a risk that stakeholders will make inappropriate decisions, with potentially damaging consequences for living resources. Information needs analysis is the process whereby needs expressed in a variety of ways — narrow, broad, technical or bureaucratic — are guided into a consistent, mutually agreed, set of information objectives. A variety of tools and methods exist to facilitate and structure the analysis process.

Further reading: Volume 2.

3. Design information products and services

In reality, many groups of decision-maker are too busy or lack the technical background to process large amounts of data or apply themselves to difficult interpretation tasks. They need brief summaries of complex issues, presented in such a way that they can be absorbed quickly without the need for special tools or expertise. By emphasising presentation issues such as clarity, timing and method of delivery, information products render information useful and usable by its intended audience. The aim is to take account of the constraints under which users operate, and tailor information accordingly. This is why the results are called information *products* rather than information *sources*, reinforcing the idea that they provide a specific service to their users.

Further reading: Volume 3.

4. Agree roles and responsibilities

The information needed to conserve and sustainably use living resources is multi-disciplinary in nature, even when confined to a single sector such as forestry or agriculture, and may be required on a dynamic and variable set of topics. The datasets needed to generate this information may be scattered amongst many organisations and sources, making the task of integration and interpretation complex. A further challenge arises when some organisations are unable or unwilling to provide access to their data.

Information networks, which are simply assemblies of individuals, groups and organisations with common information objectives, address this challenge by focusing on cooperation. The aim is to build trust and confidence between network partners, who may include scientists, policy-makers and resource managers, leading to improved uptake of scientific information in policy and planning. The success of a network depends on its partners understanding how they are expected to contribute to the network's goals and, equally importantly, how they stand to benefit from its existence. Their roles and responsibilities for information production should be defined clearly and reviewed as necessary.

Further reading: Volume 4 [optional Volume 5].

5. Build capacity

Constraints in information management capacity can impede progress towards corporate or network objectives. Clear priorities for capacity building are needed, and the greatest challenge is deciding how and where to focus investments. The latter should, wherever possible, be based on an assessment of where existing capacities are located and how readily these can be mobilised for specific tasks. This can be achieved by surveying the range of data, skills and facilities which are available, and the management systems which bind these together.

Typically, some information objectives will be achieved simply by improving coordination between organisations and by sharing scarce resources. Others may require direct financial support to realise. It is the role of senior managers in the organisations concerned to attempt to enable progress in both cases. The aim is to extract the greatest benefit from existing resources as a first step, but to facilitate access to additional resources as required for specific tasks.

Further reading: Volume 6.

9 CASE STUDIES

9.1 Overview

Since the Earth Summit a number of publications have begun to address the challenge of generating information in support of the goals of the Convention on Biological Diversity. These include the *Guidelines for Country Studies on Biological Diversity* (UNEP 1993) which provides advice on assessing the conservation status and economic potential of living resources, *Guidelines on National Biodiversity Planning* (Miller and Lanou 1995), which examines how to develop biodiversity strategies, action plans and reports, and the more specific *Guide to Information Management* (UNEP/WCMC 1996) and *Guide to National Institutional Survey* (UNEP/WCMC 1998) which illustrate the steps necessary to build a national infrastructure for biodiversity information management.

Two fundamental lessons for information management can be drawn from the experiences captured by these publications, and from related projects and activities which have addressed the issue of managing information on living resources following the Earth Summit:

1 Information should be **relevant to policy and management goals**, whether these apply at the local, national or international levels. It follows that there is a need to find ways of converting the abundant information on environmental, economic and social issues into forms suitable for decision-making.

2. Information should reflect the real complexity of environmental concerns by involving an **appropriate range of stakeholders** in the information production process.

In accordance with these principles, many governments in both North and South have developed strategies for improving the accessibility, quality and relevance of national biodiversity information. Two examples follow.

9.2 The UK Biodiversity Database

Following the Earth Summit the United Kingdom prepared a biodiversity action plan in accordance with Article 6a of the Convention on Biological Diversity. One of the key strengths of the plan, which was published in January 1994 (UK Government 1994), is that it draws together existing policies and programmes for nature conservation throughout the country.

To advise the UK Government on implementation of the Plan, a Biodiversity Steering Group was established with representation from key sectors and nature conservation institutions. The primary tasks of the Steering Group, which was chaired by the Department of Environment, were to develop costed targets for conserving threatened, declining or otherwise important species and habitats; to suggest ways of improving accessibility and coordination of biodiversity information; to recommend ways of increasing public participation and awareness of the need to conserve biodiversity; and to monitor and oversee implementation of the Plan.

The United Kingdom has a rich reserve of biodiversity data resulting from the efforts of dedicated amateurs and government agencies. However, much of this data is not available to decision-makers in a form which is relevant to their needs. There are also important gaps and overall coordination of data collection and management activities is lacking. For instance, policy-makers may be more interested in the reasons why species or habitats are disappearing than their rates of loss (i.e. information on pressures as well as status). This may demand a programme of monitoring and research drawing on a baseline of consistently structured data, which may not exist at the present time, despite the wealth of biodiversity data which have been collected in support of earlier, often uncoordinated priorities.

To make more effective use of what data already exist, and to coordinate activities in future, the Biodiversity Steering Group established three information management priorities: to improve access and coordination of existing biodiversity datasets; to provide common standards for future biological recording; and, most ambitiously, to examine the feasibility of a UK Biodiversity Database (UKBD). Planning was driven by the knowledge that the data requirements of most users are actually quite similar and that, as a consequence, the most efficient systems and mechanisms were those which resulted in data being recorded, checked and stored only once, but accessed and used many times for many purposes.

Emphasis was placed on prioritisation of datasets for decision-making processes, the management of data quality (validation, security and documentation), and the preparation of data exchange policies. It was recommended that major datasets should be managed to certain standards of content, quality and accessibility by an agreed custodian, and made available to other users for legitimate purposes. This is in direct agreement with European information regulations on freedom of access to information on the environment, which state that public sector holders of data have an obligation to provide access to and ensure the quality of their data.

When considering whether (and how) to make datasets available to external audiences, the twin issues of charging for data and protection of intellectual property rights (IPR) are often cited by institutions and individuals as potential barriers. This need not necessarily be so if the right environment for data exchange is created at an appropriately high level, and the characteristics of good data exchange policies are agreed. For instance, the UK Biodiversity Steering Group recognises that different charges and conditions of use of data might be issued to different classes of user as, indeed, is already common practice amongst some of the institutions represented.

With regard to standards, three thematic areas were identified: biodiversity standards covering the data content (e.g. species nomenclature, vegetation classifications, threat categories); information standards covering data structures (e.g. the attributes recorded for a species distribution record); and technology standards covering the compatibility and interoperability of information and communications technology. The task of reviewing existing standards and addressing weaknesses with new standards has begun, on the premise that data will continue to be stored on a variety of media (e.g. paper, spreadsheet, database) and communicated in a variety of ways (e.g. by hand, post, CD-ROM, Internet).

Clearly, the most significant component of the Biodiversity Steering Group's information strategy is the UKBD. When established, this will comprise a network of cooperating partners working towards agreed standards of data quality and access. The UKBD will operate at two levels: the local level, where data are commissioned primarily for the development of local biodiversity action plans under the management of a consortium of competent local organisations; and the national level, where cooperation will be largely voluntary and will be facilitated by one body, most probably the Joint Nature Conservation Committee (JNCC). This approach encourages the provision of relevant information to local planning processes, but also enables integration of high-quality local- and national-level datasets for national biodiversity planning.

With so many datasets currently in existence or under development, an important first stage of the UKBD is a stock-taking exercise, or dataset catalogue. Presented in various forms, from a simple text directory to an electronic metadatabase, the catalogue will document what datasets are available, assess their quality and accessibility, enable major gaps to be determined, and help to reduce duplication of effort and redundancy.

[Source: UK Government 1995]

9.3 The Indira Gandhi Conservation Monitoring Centre

India is one of the twelve 'megadiversity' countries which, together, possess 60-70% of the world's total biodiversity. Its ten biogeographic zones subsume a very wide range of ecosystems. India has about 7% of the world's flowering plant species, 14% of the world's bird species and, overall, 81,000 species of animals representing 6.4% of the world's identified fauna. Further, one third of its 15,000 flowering plants are endemic to India, plus 14% of its 1,228 bird species, 32% of 446 reptile species, and 62% of its 204 amphibians. The marine habitat covers 7,500 km of coastline extending 200 nautical miles off-shore into its Exclusive Economic Zone.

Presently, biodiversity data are held unevenly across the country in different, often incompatible formats. Information based on these data is urgently required by decision-makers and other users for different purposes, such as protected areas management, environmental assessment, land-use planning and development, awareness-raising and education, research, and prioritisation of conservation activities. A particularly important need is access to reliable data on the socio-economic factors affecting the potential for conservation and sustainable use of living resources in the regions around protected areas. Although many institutions collect data useful for biodiversity conservation (some are mandated to distribute this), data are rarely made accessible in reasonable time or in the formats users require, particularly where the data transcend institutional jurisdictions. This is because data are primarily collected for internal use by the institutions concerned, and there is little awareness of the need to build up essential datasets for common, nationally-agreed goals.

The Indira Gandhi Conservation Monitoring Centre (IGCMC) was set up by the World Wide Fund for Nature (WWF-India) in May 1994, with the support of the Ministry of Environment and Forests, Government of India, to support the conservation of biodiversity and natural resources in India through the provision of scientific and management information. IGCMC aims to support conservation by helping to coordinate the management and dissemination of information on India's living resources, the pressures they most acutely face, and on the performance of government and private sector policies on land use, resource management (e.g. water, forests, wildlife) and protected areas. In cooperation with a series of major data management partners, the agreed mission of IGCMC is as follows:

"*to support biodiversity and natural resources conservation in India through collecting, managing, disseminating and making accessible relevant data and knowledge, and by providing appropriate technical, analytical and networking services*"

Considering IGCMC's priority for producing biodiversity information and its current emphasis on protected areas, its short-term objectives are to gather, interpret and distribute information on threatened plants and animal species, trade in wildlife, wetlands, eco-development planning in and around protected areas, and the distribution and status of habitats and species in the Eastern Himalaya, Western Ghats and Andaman and Nicobar Islands.[2] The longer-term strategy is to provide more comprehensive information and capacity building services to a wider range of users, notably in the private sector, by consolidating existing data-management partnerships and developing new ones. IGCMC is implementing this strategy with support from the British Department for International Development (formerly Overseas Development Administration).

The sustainable management of natural resources depends on policies which reflect a wide variety of interests. The development of such policies depends upon a similarly wide base of data and information. One task of IGCMC is to increase access to key datasets via the promotion of common approaches (standards) for data management and exchange. This demands excellent coordination between IGCMC and its data-management partners, which can be facilitated by electronic networking. IGCMC has developed a model for the integration of diverse and distributed datasets which makes extensive use of Internet-based communication and presentation tools. These enable the growing network of data management partners to cooperate in the development and delivery of information to specific groups of users. No advanced technology will be required by users other than Internet connectivity, which is growing rapidly in India with significant governmental support.

---

2 Under the 'Biodiversity Hotspots' Conservation Programme of WWF-India.

IGCMC is set to play an important role in providing information to decision-makers in many sectors and at many levels in India. One example is the provision of objective information to environmental appraisal committees in the Ministry of Environment and Forests. The role of these committees is to review environmental impact assessments (EIA) for major development projects, and approve or reject the projects as necessary. IGCMC will also help the Government of India fulfil its reporting obligations under the Convention on Biological Diversity. Specifically, it will assist in the following areas: preparation of a national biodiversity conservation strategy (Article 6), identification and monitoring (Article 7), environmental assessment (Article 7), forecasting (Articles 7 and 10), *in-situ* conservation/sustainable development (Articles 8 and 10), *ex-situ* conservation (Article 9), research and training (Article 12), public education and awareness (Article 13), exchange of information (Article 12), technical and scientific cooperation (Article 18), and scientific and technical advisory committee (Article 25).

[Source: S.K. Puri, Indira Gandhi Conservation Monitoring Centre, WWF-India, New Delhi]

10 REFERENCES

Brookes, C.H., Grouse, P.J., Jeffery, D.R. and Lawrence, M.J. 1982. *Information System Design*. Prentice-Hall.

Miller, K.R. and Lanou, S.M. 1995. *National Biodiversity Planning: Guidelines Based on Early Experiences Around the World*. World Resources Institute, United Nations Environment Programme, and IUCN – The World Conservation Union. Washington D.C.; Nairobi; Gland, Switzerland.

Newall, A., Shaw, J.C. and Simon, H. 1958. Elements of a Theory of Human Problem Solving. *Psch. Review*, May 1958, pp 151-166.

Sandbrook, R. 1994. *Annual Report 1994–1995*. International Institute for Environment and Development, London, UK.

UK Government 1994. *Biodiversity: The UK Action Plan*. The Stationery Office Ltd (HMSO), London, UK.

UK Government 1995. *Biodiversity: The UK Steering Group Report. Volume 1: Meeting the Rio Challenge*. The Stationery Office Ltd (HMSO), London, UK.

UNEP 1993. *Guidelines for Country Studies on Biological Diversity*. United Nations Environment Programme, Nairobi, Kenya.

UNEP/WCMC 1996. *Guide to Information Management in the Context of the Convention on Biological Diversity*. United Nations Environment Programme, Nairobi, Kenya.

UNEP/WCMC 1998. *Guide to National Institutional Survey in the context of the Convention on Biological Diversity*. United Nations Environment Programme, Nairobi, Kenya.

WRI/UNDP 1990. *World Resources 1990–92*. Oxford University Press, New York, USA.